The Cow's Girl

THE MAKING OF A REAL COWGIRL

by CHARLOTTE CALDWELL

ISBN 10: 1-59152-148-3
ISBN 13: 978-1-59152-148-8

Published by Barn Board Press

For more information, contact Farcountry Press, PO Box 5630, Helena, MT 59602; 1-800-821-3874.

You may order extra copies of this book by calling Farcountry Press toll free at (800) 821-3874.

sweetgrassbooks
a division of Farcountry Press

Produced by Sweetgrass Books; PO Box 5630, Helena, MT 59604; (800) 821-3874; www.sweetgrassbooks.com.

Printed in the United States of America.

19 18 17 16 15 1 2 3 4 5

This book is dedicated to ranchers, the ranching way of life, and the next generation of Cowgirls.

I am forever enriched by and grateful to the Lannen family. They continuously welcome me into their lives and generously share the experiences of life on a family ranch.

This book is the sequel to *The Cow's Boy: The Making of a Real Cowboy*.

My name is Olivia. I'm a Black Angus heifer calf. When I was three days old, my momma and I accidentally got separated by a fence. I was lost and hungry. Cowgirl adopted me, and named me Olivia. She wrote my name on my ear tag. My momma's identification is on the other side of the tag, because it is important to know who my momma is.

My Cowgirl loves me. I'm her bottle-baby calf. She has decided to take me to the county fair for the 4-H Bucket Calf Project.

A bucket calf is an adopted calf raised on formula—a powdered food mixed with water. Some adopted calves suck milk from a bucket, but not me.

I'm glad Cowgirl feeds me from a bottle, so she can pat my head at the same time.

She prepares and feeds me two bottles of formula every morning and night.

Before she adopted me, she had adopted a bull calf that she named Oliver. Her little brother takes care of Oliver now.

He will take Oliver to the county fair as his 4-H Bucket Calf, too.

I like to suck on Cowgirl's fingers after I have finished my bottles. She tells me that my belly is full, but my brain still thinks I'm hungry. So I suck on her fingers until my brain catches up with my belly.

After my belly is full, I play with Cowgirl. We play follow-the-leader. I try to jump too. Then she kisses me and tells me to play with Oliver, because she has other ranch chores to do.

I'm not supposed to follow her into the house, but sometimes the door is left open and well, hmmm . . . I just can't resist.

My Cowgirl also takes care of dogs, cats, chickens, and her bunny rabbit, Paisley.

She has two itty-bitty kitties that need to be fed, too. Their bottle is really itty-bitty! Momma cats lick their babies clean, but Cowgirl does not lick them, she bathes them in the sink and gives them a blow dry.

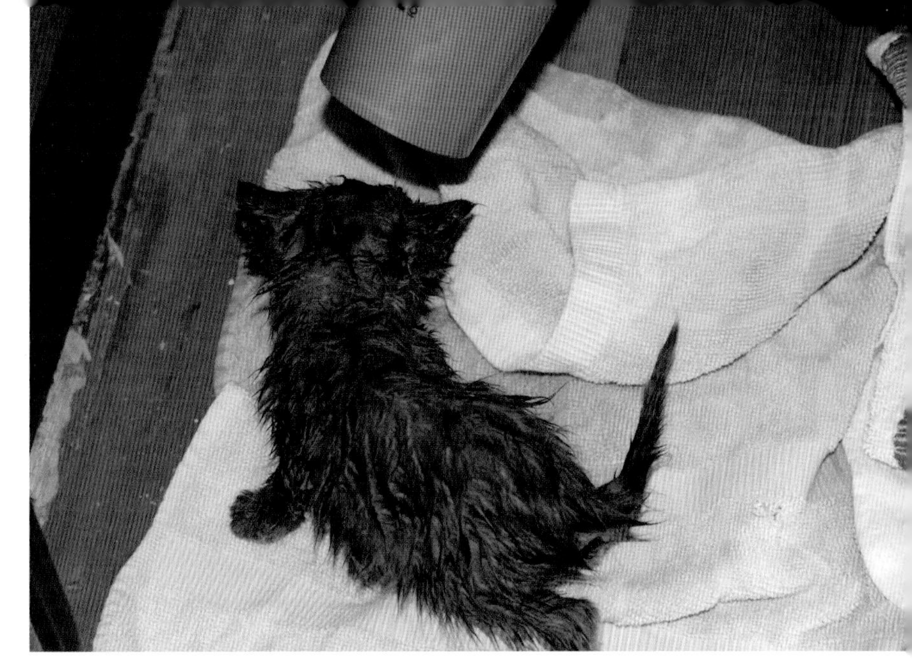

Some days, my Cowgirl and her brothers move cows to the corrals closer to the house. She grabs a halter from the barn before going to get her horse, Sassy. All the animals love my Cowgirl and come for a pat, even Jimmy the one-horned goat.

First, Cowgirl grooms her horse by brushing the dirt off her coat and by cleaning her hooves.

Then, she saddles and bridles her horse, mounts and rides off with her older brother.

She and her brother find the cows in the upper pasture, and move them down to the barn.

Sometimes, when Cowgirl's work is done, she takes a break and races with her older brother. They are good friends and have fun.

Other times, she likes to play with me, or ride her motorcycle around the ranch. She remembers to always close the gates behind her.

Cows need to be doctored, too. Cowgirl helps her daddy give the cows a vaccination to prevent diseases. They tattoo one ear of each cow to record the vaccination date.

Her momma helps her figure out the right amount of medicine to give a sick cow.

Sometimes Cowgirl tricks her older brother and locks him in the cattle chute. They laugh a lot.

When the patches on a cow's rump turn bright pink, it is time to be bred. My Cowgirl listens carefully to instructions from her daddy. Her momma helps her follow directions to make sure the cow will have a baby next spring.

Equipment occasionally breaks down. If it can't be fixed right away, then maybe its time for a little fun—playing with Oliver and me, or flying through the air on a zip line with her brothers and a neighbor.

She also likes to take her dollies, Violet and Rosy, for a ride with a friend.

Cowgirl uses the 4-wheeler to bring horses down to the barn when they are in a distant pasture.

If a bunch of cows are mixed together, the only way to know which cow belongs to which rancher is by its brand.

Cowgirl has her own brand. It is a Reverse Lazy R Hanging Diamond. A brand can go on the right or left shoulder, rib or hip. Cowgirl's brand goes on the right rib.

She and her brother sort calves to be branded. She helps by opening the chute door as someone else moves a calf through the alley.

When ranchers come together to help each other with branding, doctoring, or shipping, my Cowgirl not only helps outdoors, but she also helps with preparing the noon meal. Sometimes, she bakes a cake for all the ranchers.

After doing her homework, she comes to feed me. She doesn't have to look too hard for me, because I'm usually standing on the porch steps waiting for her. I'm hungry!

My Cowgirl and her family run a little camp for some "wanna-be-cowgirls," who live in a city. The campers learn to catch the horses, groom and tack them, and wash them after riding.

They also bottle feed the kitties, Oliver, and me.

They love doing chores with my Cowgirl!

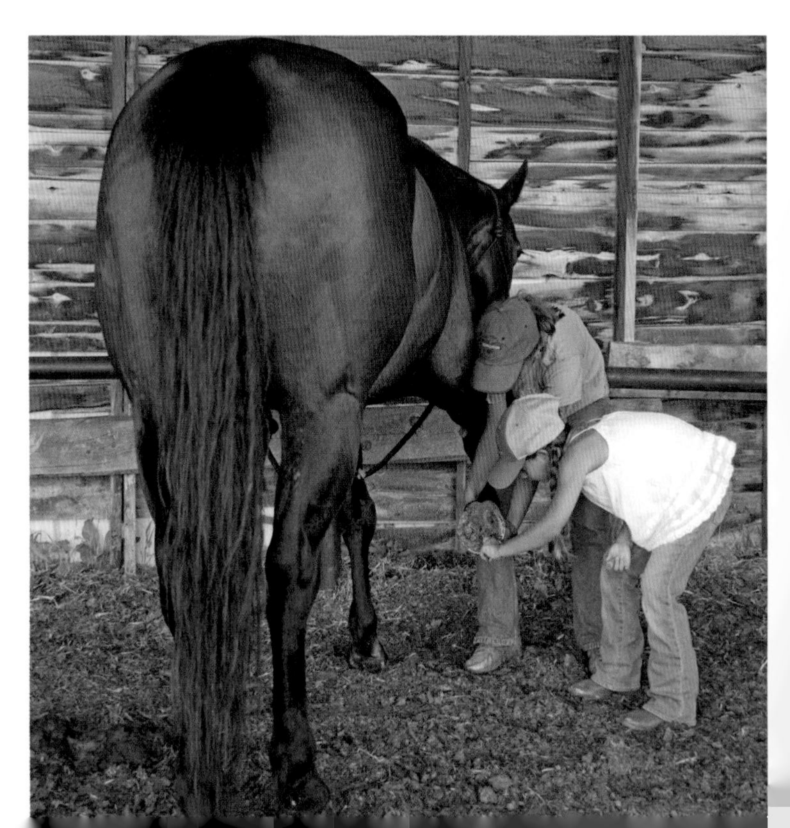

This is Abner, the 4-H steer she raised and sold in last year's county fair. He loved Cowgirl, too. She misses him.

Abner and her other 4-H steer, Humphry, were sold after the fair.

With the money she made, Cowgirl was able to pay all of her 4-H steers' expenses—their feed, salt, vitamins, and vaccinations.

She had enough money left over to buy this saddle,

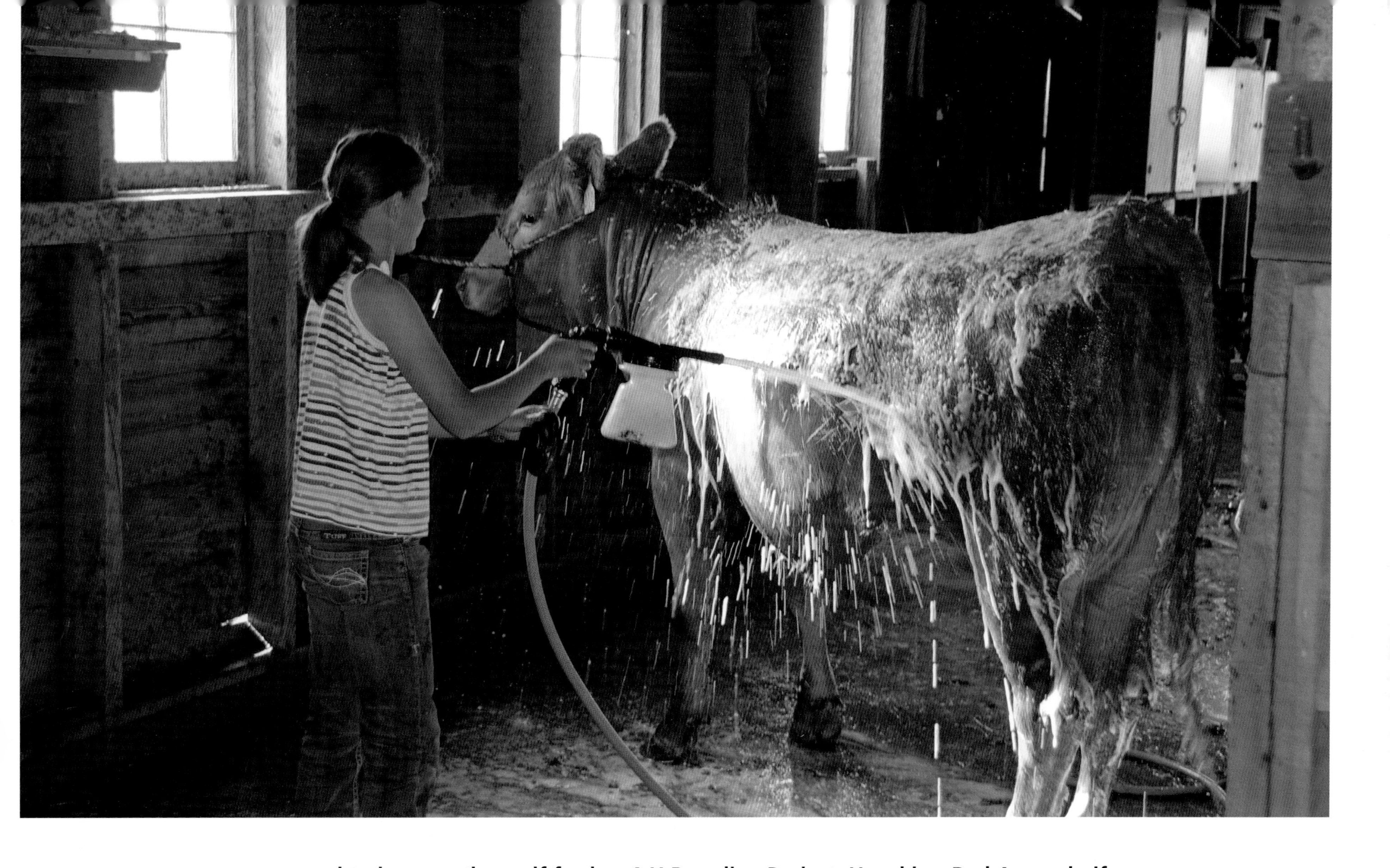

and to buy another calf for her 4-H Breeding Project. Hazel is a Red Angus heifer calf. A girl cow is called a heifer until she gives birth to a calf of her own. Cowgirl trained Hazel to wear a halter. They practice walking together for the county fair. Hazel gets washed and loved by Cowgirl.

Part of the 4-H Breeding Project is record keeping. Cowgirl writes down all her expenses—how much it costs to buy her calves, to feed them and to doctor them.

She also had enough money to buy Willis, her Black Angus steer. She entered him in the 4-H Market Beef Project.

Willis loves being shampooed. Then he gets blow-dried and clipped to look really pretty for the show.

Cowgirl wants Willis to look handsome and to weigh a lot. She walks him onto the scales and her daddy checks his weight.

Oh no, now it is our turn!

This is part of our 4-H Bucket Calf training. Oliver and I get shampooed, blow-dried and fluffed.

We like it best when Cowgirl and her brother take off our halters and let us go outdoors to munch grass. Enough of that froo froo! We'd rather moo moo!

Cowgirl practices with her steer, Willis, for months. She loves Willis and he loves her back. When she pats him, he likes to lick her. At the fair he weighed 1187 pounds and won a purple ribbon in the 4-H Market Class.

Oliver and I got purple ribbons too . . . probably because we are such cute bottle-babies!

Cowgirl was proud of her 4-H animals.

Hazel won two blue ribbons—one for 4-H Breeding Heifer class and another for Open Class.

Cowgirl's record keeping was judged too. She received Reserved Grand Champion for her record book.

My Cowgirl is a blue-ribbon girl, too. She loves Oliver and me, and all the animals on our ranch.

She also does nice things for people. She cut her hair 11 inches so she can donate it to Locks of Love, an organization for children with hair loss.

There is still a lot of work to be done on the ranch after the fair. The alfalfa hay is cut into rows and when it is almost dry, Cowgirl's daddy teaches her how to rake the two rows into one. He tells her to drive the tractor at her own pace and to stay between the two rows.

In no time, she feels confident and rakes the rest of the field by herself. Her daddy is proud of her!

Cowgirl's family takes good care of their animals. They also make sure their land is kept healthy by rotating the pastures and moving their cattle.

Earlier in the summer, Cowgirl and her family moved the cattle up to their mountain pasture to give their home place a break from grazing.

Oliver and I got to stay home, because we are the bottle-babies, and we need Cowgirl and her brother to feed us.

In the fall, it is time to trail the cattle back down to the ranch. Cowgirl and her daddy trailer their horses up to the mountain pasture. Then, they ride into the forested hillsides looking for the cows. It is a special time for Cowgirl and her daddy. They enjoy being together.

They push the cattle down the road toward the ranch. The dog helps keep the cattle moving forward together.

My Cowgirl never forgets to take care of Oliver and me.

She loves her bottle-babies, and we love her, too. She is my real Cowgirl!

Special thanks goes to my husband, Jeffrey Schutz, and my dear friend, Betsy Bishop, for reading, re-reading and editing. I'm especially grateful to Kathy Springmeyer and Shirley Machonis at Farcountry Press and Sweetgrass Books for their keen attention to detail, design and layout, and making this book happen.

4-H is the youth development program of the Cooperative Extension Service in partnership with USDA. No endorsement of this book is granted or implied by 4-H. Learn more about 4-H at www.4-h.org.

Locks of Love is a public non-profit organization that provides hairpieces to financially disadvantaged children under age 21 suffering from long-term medical hair loss from any diagnosis. The organization meets a unique need for children by using donated hair to create the highest quality hair prosthetics.

See locksoflove.com for information and donation guidelines.

Olivia and Cowgirl are loved by all the ranch animals. They like to follow Olivia and Cowgirl around.

Can you find any of their friends in the story?

Belle, the retired cattle dog

Tink, the blue heeler cattle dog

Pixie, Cowgirl's pet

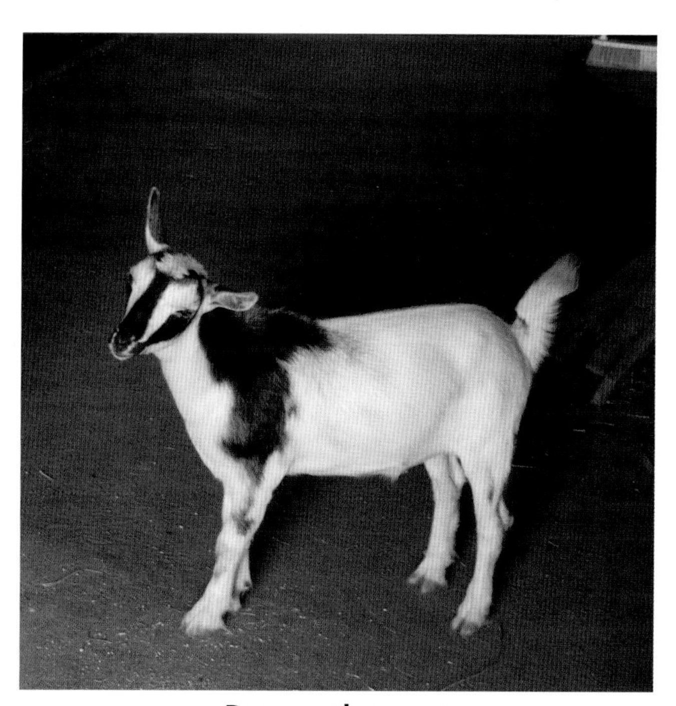

Poppy, the goat

Charlotte and her husband, Jeffrey Schutz, divide their time between their ranch in Montana and their home in Charleston, SC. This book is the sequel to *The Cow's Boy: The Making of A Real Cowboy* © 2013.

To learn more, please visit www.CharlotteCaldwell.com.